NACHIDA HAMIDI-SITOUAH, MD

DIABETES for PRIMARY CARE

A Step-by-Step Approach

ISBN: 978-1-4834-1905-3 (sc)
ISBN: 978-1-4834-1906-0 (e)

Permission for text used in this book granted by
AMERICAN DIABETES ASSOCIATION

Because of the dynamic nature of the Internet, any web addresses or links contained in this book may have changed since publication and may no longer be valid. The views expressed in this work are solely those of the author and do not necessarily reflect the views of the publisher, and the publisher hereby disclaims any responsibility for them.

Any people depicted in stock imagery provided by Thinkstock are models, and such images are being used for illustrative purposes only. Certain stock imagery © Thinkstock.

Lulu Publishing Services rev. date: 10/27/2014

Contents

1. Metformin
2. Sulfonylureas
3. The Meglitinides
4. Thiazolidinediones (TZDs)
5. Drugs Focused on the Incretin System: GLP1 Receptor Agonists and DDP4 Inhibitors
   GLP-1 receptor agonists
   DPP4 inhibitors
6. Sodium Glucose Cotransporter 2 (SGLT2) Inhibitors
7. Insulins
   A. Insulin types
      -Pharmacokinetics of SQ insulin in patients with normal kidney function
      -Premixed insulin
      -Glargine (Lantus) versus Detemir (Levemir)
      -Rapid-versus short-acting insulin (Novolog, Humalog, or Apidra vs. Regular insulin)

-U-500 regular insulin (500 units/ml)
B. Insulin Regimens
-Conventional therapy using premixed insulin (70/30, 75/25, 50/50 insulin)
-Modified fixed-dose insulin regimen
-Flexible regimen/intensive insulin therapy ("basal-bolus regimen")

1. Diabetes Pre-Visit Checklist
2. Topics to Discuss
   -Diet
   -Metformin
   -Blood-sugar monitoring
   -Targets
   -Noncompliance
      Lack of motivation
      Ignorance
      Inconvenience/Forgetfulness
      Affordability issues
      Fear from hypoglycemia
   -Ruling out inadequate insulin use
   -Injection techniques
   -Insulin storage
3. Calculating the Insulin Total Daily Dose (TDD)
4. A1c and Blood-Sugar Log Review with Insulin Adjustment
   -Hypoglycemia
      Hypoglycemia definition
      Hypoglycemia causes
      Hypoglycemia treatment
   -Hyperglycemia
      Fasting hyperglycemia

Non-fasting hyperglycemia (preprandial and/or bedtime hyperglycemia)
5. Supplemental (Correctional) Insulin
6. Insulin Dose Adjustment at Home
7. Potential Causes of Treatment Failure in Diabetes

1. Extreme Insulin Resistance
2. Insulin Pump Therapy
3. Gastroparesis
4. Obstructive Sleep Apnea
5. Double Diabetes
6. Pancreoprivic Diabetes
7. LADA

Case 1
Case 2
Case 3
Case 4
Case 5
Case 6
Case 7
Case 8
Case 9
Case 10
Case 11

Acknowledgments

This book couldn't have been completed without the help of some very important and special people, including:

Fiona Cook, MD, Clinical Associate Professor, who deserves special recognition because of her great contribution in finalizing this book (editing, positive criticism, and ideas).

Robert Tanenberg, MD, Professor of Medicine, Director of Diabetes Fellowship, for being my mentor and principal teacher during my fellowship.

My family, for all their support during this process.

Introduction

I worked for three years as a primary care physician in a community clinic. Throughout my time there, diabetes quickly became the most prevalent disease in my caseload. I was faced with the daily challenge of managing patients with uncontrolled diabetes. The length of routine follow-up visits (fifteen minutes) and the need to address other problems as well made it even more challenging. The emergence of numerous antidiabetic drugs like DPP4 inhibitors and GLP-1 receptor agonists and the complexity of some of my patients held an interest for me in pursuing a diabetes fellowship. I had no experience with the aforementioned drugs and did not feel comfortable prescribing them. Patients with "brittle diabetes" presented with significantly variable blood-sugar levels due to a strong sensitivity to food and insulin. Finding the right insulin dose for them seemed nearly impossible, so I had to refer them to a subspecialist. I also had many patients who were noncompliant with the diet, medications, and insulin. Nothing I did or said seemed to make a difference, but I didn't expect a better knowledge in the field to help me manage them differently. In fact, in the hospital where I did my residency, the endocrinologists used to instruct us not to refer noncompliant patients to them because it was considered as a waste of expertise. We were advised to refer them to diabetic educators instead.

After I completed my fellowship, I returned to my previous job at the same community clinic and mostly started seeing patients with uncontrolled diabetes. I was surprised to realize that what I learned the most from this fellowship was how to better use the old drugs, especially insulin, rather than how to use the new ones. I did learn about new drugs as well as insulin pumps, insulin U-500 regular, and CGM (continuous glucose monitoring) interpretation, but these tools are only used for a minority of the patients with diabetes. I left the fellowship with a simple conclusion that by spending time with the patients and taking a good history, it becomes possible to improve diabetes control. The solution is an open and honest communication with the patient. This is the essence of the patient-centered approach, which means to not focus on how to make the patients follow our recommendations but to focus instead on what they are doing and how they are doing it.

After making this discovery, the idea of sharing it with others excited me. Diabetes is managed mainly by primary care physicians, who seek and require more expertise. I have made many mistakes that my fellowship has taught me to fix, so I decided to write this book in an effort to keep others from making the same errors. This is not a book about medical guidelines. This is a book about a process of analytical thinking that will lead you to discover the most common causes of treatment failure in patients with diabetes. It will list some of the frequent oversights made by primary care physicians and will show how experts approach diabetes. I like to compare it to a cookbook wherein you will not find the exact ingredients of a recipe but the steps and tips that will help you succeed in your own method. We focus on multiple daily insulin injections ("basal-bolus" regimen).

There are many reasons for treatment failure, and finding these reasons in each patient involves an elimination process. The process can be likened to taking a multiple-choice exam: in order to find the right answer, you first need to eliminate all the wrong ones.

In managing diabetes, finding what went wrong is the most important step. There are many possible reasons for poor glycemic control, more than simply an insufficient dosage or a noncompliant patient. With each case, you need to ask yourself questions toward finding what is wrong and fixing it. Even after finding problems and offering solutions, you will need to readdress them more than once until you obtain a confirmation of their resolution.

Part 1

Overview of Most Commonly Used Antidiabetic Agents

1. Metformin

Metformin is an insulin sensitizer, and it remains the first choice and the cornerstone of therapy in type 2 diabetes (T2DM) because of its efficacy and safety. The maximal dose is 1000mg bid with meals.

2. Sulfonylureas

This class of drugs has fallen out of favor because of their association with weight gain, hypoglycemia, and limited duration of effectiveness. However, along with metformin, they do remain the least expensive and most potent agents in early diabetes.

"Glipizide should be the sulfonylurea of choice in patients with advanced kidney dysfunction. Caution should be used with the extended-release form". [1]

---

[1]   Reilly JB, Berns JS. Selection and dosing of medications for management of diabetes in patients with advanced kidney disease. Seminar in dialysis-Vol 23,No 2 (March-April) 2010 pp. 163-168.

3. The Meglitinides

Repaglinide (Prandin) and Nateglinide (Starlix).

Meglitinides are shorter-acting insulin secretagogues with less hypoglycemia than sulfonylureas. Consider using them in patients with irregular meal schedules, or patients who skip meals or develop hypoglycemia on sulfonylureas.

4. Thiazolidinediones (TZDs)

Pioglitazone appeared to have a modest benefit on cardiovascular events in one large trial; it has been associated with a possible increased risk of bladder cancer. Avoid it in patients who are at risk for bladder cancer, including smokers.

5. Drugs Focused on the Incretin System: GLP-1 Receptor Agonists and DPP4 Inhibitors

Earlier and more frequent use of incretin-based therapy is now recommended.

GLP-1 Receptor Agonists

Exenatide (Byetta/Bydureon) given bid and once a week respectively and Liraglutide (Victoza) given once a day.

Action: These drugs stimulate pancreatic insulin secretion in a glucose-dependent fashion, suppress pancreatic glucagon output, slow gastric emptying, and decrease appetite.

They are associated with modest weight loss.

These drugs lower Hb Alc levels by about 1 percent and do not cause hypoglycemia given alone or in combination with metformin.

Side effects/caution: nausea, vomiting and diarrhea, increased risk of pancreatitis (rare). Therapy "should be discontinued promptly in patients with abdominal pain until pancreatitis is ruled out as a cause".[2]

These drugs should be avoided in patients with chronic kidney disease stage 3-5 and in patients with personal history of thyroid cancer, unevaluated thyroid nodules, or with family history of medullary thyroid CA or MEN-2.

Best candidates: patients with T2DM, obese/overweight, at risk for hypoglycemia, recent onset DM, with good medical insurance and willing to take injections.

DPP4 Inhibitors

Sitagliptin (Januvia), Linagliptin (Tradjenta), Saxagliptin (Onglyza), Vildagliptin (Galvus), and Alogliptin (Nesina).

Action: "Enhance circulating concentrations of active GLP-1 and GIP. Their major effect appears to be in the regulation of insulin and glucagon secretion; they are weight-neutral".[3] They do not cause hypoglycemia by themselves.

Best candidates: patients with insurance, with postprandial hyperglycemia, at risk for hypoglycemia, recent onset of diabetes.

The use of the combination of GLP-1 agonist and DDP4 inhibitor is contraindicated.

---

[2]   Schwartz SS, Kohl BA. Glycemic control and weight reduction without causing hypoglycemia: the case for continued safe aggressive care of patients with type 2 DM and avoidance of therapeutic inertia. Mayo Clin Proc. 2010 Dec;85(12 Suppl):S15-26. doi: 10.4065/mcp.2010.0468. Epub 2010 Nov 24.

[3]   Inzucchi SE, Bergenstal RM, Buse JB, Diamant M, Ferrannini E, Nauck M, Peters AL, Tsapas A, Wender R, Mathews DR. Management of hyperglycemia in type 2 diabetes: a patient-centered approach: position statement of the American Diabetes Association (ADA) and the European Association for the Study of Diabetes (EASD). Diabetes Care.2012 Jun;35(6):1364-79. doi: 10.2337/dc12-0413. Epub 2012 Apr 19.

6. Sodium Glucose Cotransporter 2 (SGLT2) Inhibitors

Canagliflozin (Invokana) and Dapagliflozin (Farxiga).

This is a new class of oral antidiabetic agents.

SGLT2 inhibitors lower the renal threshold for glucose, markedly increasing urinary glucose excretion.

These drugs seem to lower A1C by 0.45 to 0.92 percent, with an associated weight loss of 0.7 to 3.5 kg.

They should be used with caution, preferably as "add-on therapy to two or three other agents, including insulin, in patients who would benefit from weight loss".[4]

Their primary side effects are increased urinary tract and genital infections and dehydration; however, an unexplained adverse effect is increased LDL-C.

They should be avoided in patients with renal failure and are contraindicated in severe renal impairment (<30 mL/minute/1.73 m^2) and ESRD.

---

[4]   Garber AJ, Abrahamson MJ, Barzilay JI, Blonde L, Bloomgarden ZT, Bush MA, Dagogo-Jack S, Davidson MB, Einhorn D, Garvey WT, Grunberger G, Handelsman Y, Hirsch IB, Jellinger PS, McGill JB, Mechanick JI, Rosenblit PD, Umpierrez GE, Davidson MH. American Association of Clinical Endocrinologists' comprehensive diabetes management algorithm 2013 consensus statement. Endocr Pract. 2013 May-Jun;19(suppl 2).

7. Insulins

A. Insulin Types

Pharmacokinetics of SQ insulin in patients with normal kidney function

Insulin	Onset of Action	Peak of Action	Duration of Action
Aspart (Novolog), lispro (Humalog), or glulisine (Apidra)	10-20 min	40-50 min	3-5 hrs
Human Regular (Humulin R, Novolin R)	30-60 min	1-5 hrs	4-12 hrs
Human NPH (Humulin N, Novolin N)	1-2 hrs	4-14 hrs	10-24 hrs
Glargine (Lantus)	1.1 hrs	Peakless	About 24 hrs
Detemir (Levemir)	1.1-2 hrs	Relatively flat	Dose-dep 5.7-24 hr
U-500 Regular	30 min	1-3 hrs	12-24 hrs

Premixed insulin

Humulin 70/30 or Novolin 70/30	70% NPH	30% Regular
Human 50/50	50% NPH	50% Regular
Humalog 75/25	75% Lispro protamine	25% Lispro (Humalog)
Novolog 70/30	70% Aspart protamine	30% Aspart (Novolog)

Glargine (Lantus) versus Detemir (Levemir)

These are both "basal" insulins, typically given at twenty-four-hour intervals and designed to cover patients' insulin needs in between meals and overnight.

Levemir has the advantage of being weight-neutral, compared to Lantus, which is associated with some weight gain, but Levemir has a shorter duration of action.

If used in T1DM or patients without endogenous insulin, it is best to split Levemir into two doses twelve hours apart.

"Most comparative trials [show] a higher average unit requirement with insulin Detemir"[5] compared to Glargine.

Rapid- versus short-acting insulin (Novolog, Humalog, or Apidra vs. Regular insulin)

These are "bolus" or "prandial" insulins used for mealtime coverage. Rapid-acting insulins have a more physiologic action and are generally preferred if affordable.

Unlike regular insulin, they have a faster onset and a shorter duration and their time to peak action is independent of the dose.

The use of regular insulin may be preferred in patients who snack between meals or in patients with gastroparesis.

U-500 regular insulin (500 units/ml)

This "is an insulin formulation that is five times more concentrated than the usual U-100 insulin"[6] (100 units/ml).

"U-500 is a human regular insulin, but its time to action is not the same as the usual U-100 formulation. It has the same peak as U-100

---

[5]   Inzucchi SE, Bergenstal RM, Buse JB, Diamant M, Ferrannini E, Nauck M, Peters AL, Tsapas A, Wender R, Mathews DR. Management of hyperglycemia in type 2 diabetes: a patient-centered approach: position statement of the American Diabetes Association (ADA) and the European Association for the Study of Diabetes (EASD). Diabetes Care.2012 Jun;35(6):1364-79. doi: 10.2337/dc12-0413. Epub 2012 Apr 19.

[6]   Valentine V., CNS, BC-ADM, CDE, FAADE. Don't resist using U-500 insulin and pramlintide for severe insulin resistance. Clinical Diabetes; Spring2012, Vol 30: 80-84.

regular insulin, but its duration is more like that of NPH insulin, [about twelve] hours"[7] or even as long as twenty-four hours in some patients.

U-500 is usually indicated when the total daily dose of U-100 insulin reaches 200-300 units/day. It is important to determine that the patient is actually taking the prescribed doses before making a change to U-500.

B. Insulin Regimens

Conventional therapy using premixed insulin (70/30, 75/25, 50/50 insulin)

This may be convenient, as premixed insulins are generally administered twice a day, but is inflexible due to the inability to titrate the shorter-acting insulin separately from the longer acting insulin. The result is more hypoglycemia and weight gain and less-optimal glycemic control than the multiple daily injections. Premixed insulins should be reserved for patients who eat regularly and cannot use intensive insulin therapy. They should be avoided in patients with type 1 diabetes (T1DM) or with severe insulin deficiency and in patients who have a tendency to miss meals or who have unpredictable eating habits. While skipping a meal, these patients would either miss the insulin dose, resulting in high blood glucose from lack of basal insulin for many hours, or take the dose, resulting in low blood glucose.

Modified fixed-dose insulin regimen

This "refers to the use of prandial and NPH insulin in the morning before breakfast, prandial insulin before dinner, and intermediate- or long-acting insulin at bedtime".[8]

---

[7]   Valentine V., CNS, BC-ADM, CDE, FAADE. Don't resist using U-500 insulin and pramlintide for severe insulin resistance. Clinical Diabetes; Spring2012, Vol 30: 80-84.

[8]   Mehta SN, Wolfsdorf JI. Contemporary management of patients with type 1 diabetes. Endocrinol Metab Clin North Am. 2010 Sep;39(3):573-93. doi: 10.1016/j.ecl.2010.05.002.

There usually is no need for a lunchtime insulin dose due to the peak action of the morning dose of NPH.

Flexible regimen/intensive insulin therapy ("basal-bolus regimen")

This is now the gold standard for insulin therapy.

A graduated approach may be used by first adding prandial insulin before the meal with the greatest carbohydrate content, often the evening meal.

If needed, dosing may be titrated up to two, then three, mealtime insulin injections.

Part 2

Implementation Strategies of Antihyperglycemic Therapy

- The ADA now recommends individualized treatment targets: a hemoglobin A1c target of < 7% in most patients, a target of 6-6.5% in young patients without any significant cardiovascular disease, and a target of 7.5-8% in "patients with a history of severe hypoglycemia, limited life expectancy, advanced complications, [or] extensive comorbid conditions".[9]
- Start with metformin if not contraindicated.
- For T2DM on metformin and A1c > 7%, use two-drug combination therapy.
- After three months, if A1c target is not reached, advance to three-drug combination therapy, which might include basal insulin.
- If the above combination fails to achieve A1C target after three to six months, start insulin or proceed to a more complex insulin regimen.

---

[9] Inzucchi SE, Bergenstal RM, Buse JB, Diamant M, Ferrannini E, Nauck M, Peters AL, Tsapas A, Wender R, Mathews DR. Management of hyperglycemia in type 2 diabetes: a patient-centered approach: position statement of the American Diabetes Association (ADA) and the European Association for the Study of Diabetes (EASD). Diabetes Care.2012 Jun;35(6):1364-79. doi: 10.2337/dc12-0413. Epub 2012 Apr 19.

- For patients with an A1c > 9%, consider starting insulin if not done yet unless the patient is able to cut down a large amount of sugar-sweetened drinks.
- Start basal insulin at 0.1-0.25 units/kg/day for most patients. Titrate it to 0.4-0.6 units/kg/d if needed.
- For patients with an A1c > 10% and/or severe hyperglycemia, advance rapidly to multiple daily insulin doses.
- Adding prandial insulin is usually indicated once basal insulin dose reaches 0.5 units/kg, especially if it approaches 1 unit/kg if A1c remains elevated (+/-) with postprandial BG >180.
- Consider using a graduated approach when adding prandial insulin.
- Once progressed to multiple insulin doses, stop oral antidiabetic medications except for metformin and thiazolidinediones (TZDs).

However, "TZDs [may need to] be reduced in dose (or stopped) to avoid edema and excessive weight gain".[10]

- Keep the insulin basal/prandial ratio around 1(50/50) for most patients.
- Adjust prandial dose to meal carbohydrate content in patients who tend to eat very different amounts of carbohydrate at different meals.
- Keep in mind the effect of the antidiabetic therapy on the patient's weight. Insulin, sulfonylureas, and thiazolidinediones cause weight gain.

Try to keep the patients on the minimal insulin dose needed to control their diabetes. Attempts to decrease insulin dose might be made even in the absence of hypoglycemia. Patient's

---

[10]   Inzucchi SE, Bergenstal RM, Buse JB, Diamant M, Ferrannini E, Nauck M, Peters AL, Tsapas A, Wender R, Mathews DR. Management of hyperglycemia in type 2 diabetes: a patient-centered approach: position statement of the American Diabetes Association (ADA) and the European Association for the Study of Diabetes (EASD). Diabetes Care.2012 Jun;35(6):1364-79. doi: 10.2337/dc12-0413. Epub 2012 Apr 19.

endogenous insulin secretion could be suppressed by high exogenous insulin dose.

- Use supplemental insulin when needed, based on the premeal blood-sugar level.

Part 3

Practical Approach
(Office Visit)

1. Diabetes Pre-Visit Checklist

Please check all that apply.

Do you eat 3 meals per day? ... ... □ YES □ NO

Does each meal contain about the same amount of carbohydrates? . ...
... ... □ YES □ NO

-If not, check the box that applies.

BREAKFAST	**LUNCH**	**DINNER**
□ low carb	□ low carb	□ low carb
□ medium carb	□ medium carb	□ medium carb
□ high carb	□ high carb	□ high carb

Carbohydrates (carbs) are starches and sugars.

Examples of high-carb meals: meals containing a significant amount of pasta, bread, rice, potatoes, etc.

Examples of low-carb meals: meals consisting mostly of meat and vegetables.

Do you snack between meals or at bedtime? .. .. □ YES □ NO

Do you drink sweet tea, regular soda, fruit juice, or other sugary drinks?.. ...
... ... □ YES □ NO

-If you answered yes, how often do you have sugary drinks?

□ less than 1 drink per day □ 1 drink per day □ more than 1 drink per day

Are you currently taking metformin (Glucophage)? If the answer is yes, please answer the following 3 questions.

Does metformin cause you to have an upset stomach or diarrhea?□ YES □ NO

Do you take a smaller dose than the dose that is prescribed to you by your
doctor?. ... ... □ YES □ NO

Do you ever miss your metformin?. ... ... □ YES □ NO

-If you answered yes to the last question, how often do you miss your
metformin? Please check the box that applies.

□ less than 1 time per week □ 1 time per week □ 2–3 times per week □ 4–6
times per week □ 1–2 times per day

Do you currently check your blood-sugar level? ... ... □ YES □ NO

If so, how often are you checking it? _____

Do you bring your meter and/or logbook to every appointment?□ YES □ NO

-If not, you should start doing it. Your doctor cannot offer you the best care
for your diabetes without information on your blood-sugar levels.

Do you have difficulties affording the meter strips?. ... ... □ YES □ NO

Did you know that the most affordable meter and strips in the area are (...) and they cost (..$)?.. ... ... □ YES □ NO

Do you know what your target blood-sugar range is?.. ... ... □ YES □ NO

-If you did not bring your meter or logbook with you today, please answer the following question.

What is your average blood-sugar level? Please check all the boxes that apply; you can circle more than one.

Before Breakfast	Before Lunch	Before Dinner	At Bedtime
□ less than 80	□ less than 80	□ less than 80	□ less than 80
□ 80–100	□ 80–100	□ 80–100	□ 80–100
□ 100–150	□ 100–150	□ 100–150	□ 100–150
□ 150–200	□ 150–200	□ 150–200	□ 150–200
□ 200–250	□ 200–250	□ 200–250	□ 200–250
□ 250–300	□ 250–300	□ 250–300	□ 250–300
□ greater than 300	□ greater than 300	□ greater than 300	□ greater than 300

Did you know that, for a diabetic, a low blood sugar is generally considered a level less than 70? ... ... □ YES □ NO

Do you agree that a low blood-sugar reading is anything under 70? □ YES □ NO

-If you answered no, then what is a low blood sugar for you?_____

Are you having any blood sugars under 70?.. ... ... □ YES □ NO

-If you answered yes, how often does this occur? Please check the box that applies to you.

□ 1 time or less per month □ 2–3 times per month □ 1 time per week
□ 2 times per week □ more than twice per week

When do your lows happen the most? Please check the box that applies to you.

□ overnight □ morning □ noon □ afternoon □ evening □ bedtime

Do your low blood sugars happen when you eat a small meal?   □ YES □ NO

Do your low blood sugars happen when you are more active?... □ YES □ NO

Have you ever had a low blood sugar when someone else had to help treat you?
... ... □ YES □ NO

Have you ever been taken to the hospital for a low blood sugar?. □ YES□ NO

What do you do when you have a low blood sugar? _____

Do you snack at bedtime or wake up during the night to eat in order to avoid
a low blood sugar?.. ... ... □ YES □ NO

Do you overeat or snack between meals in order to avoid a low blood sugar?
... ... □ YES □ NO

If you are on insulin, please answer the following questions.

Do you take your insulin every day?... ... ... □ YES □ NO

If you are not taking your insulin every day, why not? _____

Do you think you could or should control your diabetes without insulin? □ YES □ NO

Do you skip insulin if your blood sugar is below a certain level in order to
avoid a low blood sugar? ... ... □ YES □ NO

If you answered yes, what is that level? _____

Do you have financial difficulties buying your insulin? .. .. □ YES □ NO

-If you answered yes, how often do you run out of your insulin? Please check
the box that applies.

□ less than 1–2 days in a month □ 3–4 days in a month □ 5–7 days in a month
□ 7–14 days in a month □ more than 14 days in a month

Do you try taking less than the dose prescribed to you by your doctor in order
to stretch out your insulin supply due to the cost?... ... ... □ YES □ NO

Do you ever fall asleep without taking your long-acting insulin?□ YES □ NO

How often do you miss taking your long-acting insulin? Please check the box that applies.

□ less than 1 time per week □ 1 time per week □ 2–3 times per week □ 4–6 times per week □ every day

Do you miss taking your insulin at work or when eating out?... □ YES □ NO

How often do you miss taking your mealtime insulin? Please check the box that applies.

□ less than 1 time per week □ 1 time per week □ 2–3 times per week □ 4–6 times per week □ 1 time per day □ multiple times a day

How do you take your mealtime insulin? Please check the box that applies.

□ more than 30 minutes before the meal □ less than 30 minutes before the meal □ during the meal □ less than 30 minutes after the meal □ more than 30 minutes after the meal

How much of the long-acting insulin do you take? _____

How much of the mealtime insulin do you take? _____

Does the dose vary each time?. ... ... □ YES □ NO

Does it depend on your blood-sugar reading?... ... ... □ YES □ NO

Do you follow a specific formula when taking your insulin?.. ... □ YES □ NO

-If you do follow a specific formula, how do you decide the amount of insulin to take? What formula do you follow?_____

Do you ever take your mealtime insulin when you skip a meal? □ YES □ NO

Do you take your mealtime insulin at bedtime or between meals to correct high blood-sugar readings? .. .. □ YES □ NO

-If you answered yes, how often do you do it?

□ less than 1 time per week □ 1 time per week □ 2–3 times per week □ 4–6 times per week □ every day

Did you know that you could leave most insulin pens at room temperature for up to 28 days?.. ... ... □ YES □ NO

Do you store insulin vials and unopened insulin pens in the refrigerator? □ YES □ NO

Do you leave insulin vials and/or insulin pens out in the heat? □ YES □ NO

If you use pens, do you count to 10 after you inject the insulin and before you remove the pen?.. ... ... □ YES □ NO

Do you rotate your injection sites?.. ... ... □ YES □ NO

Using the above checklist will help you save time and obtain the necessary information.

2. Topics to Discuss

Diet

Knowing the patient's eating habits is important, especially any significant variability in the mealtime carb content (low-, medium-, and high-carb meals) and the frequency of snacking on carbs.

One carb serving is equal to 15 grams of carbohydrates.

A medium-carb meal contains about 45-60 grams of carbs (3-4 servings) for women and about 60-75 grams of carbs (4-5 servings) for men.

NB: For simplicity, the term *meal size* (small, regular, and large) will be used sometimes, but it refers to carb content.

Always advise against drinking sugar-sweetened beverages; it is easy to address and could help achieve a significant improvement in glycemic control.

Metformin

Ask every patient with T2DM about metformin.

Ensure that every attempt is made to keep most of your patients on a maximum dose of metformin.

If the patient is not on metformin, start it after ruling out any absolute contraindication to use it.

If the patient is not on metformin or is noncompliant with it due to GI side effects, change it to metformin XR (generic) or Glumetza (a brand of extended-release metformin), which are better tolerated and associated with fewer or no GI side effects. Start with a low dose (500 mg daily) and go up progressively as tolerated.

Blood-Sugar Monitoring

It is important that you review the BG readings at different times of the day. Ask the patients on multiple daily injections to check their BG 3-4 times per day (generally before meals and at bedtime).

Discuss affordability issues. Advise the patients who are unable to check their BG frequently, to check it at rotating times of the day in order to obtain more data.

Instruct all patients to bring their meter and logbook to each visit. In the absence of a meter or a logbook, ask the patients about their average blood sugars during the day. This might be not totally accurate, but it will give you an approximation you could use temporarily.

Targets

Discuss individual A1c target.

Discuss individual blood-sugar targets, which are as follows for most patients:

-Preprandial BG target: 70-130
-Peak postprandial (1-2 hr post-meal) < 180
-Bedtime BG target: 100-140

Noncompliance

Ask about compliance and discuss the most common causes of noncompliance, which are:

-Lack of motivation

Discuss possible misconceptions, taboos, cultural barriers, lack of conviction, depression, and other priorities.

-Ignorance

Inform the patient about the need to use insulin in situations when the blood-sugar levels are normal or unknown. (It is better to take the insulin without checking BG than to miss it.)

Explain that eating smaller-size meals doesn't always eliminate the need for prandial insulin use.

-Inconvenience/Forgetfulness

Offer the option to switch from vials to pens and ask the patients to carry the pen with them.

For patients who are unwilling to take insulin four times a day, consider the following options:

- A graduated approach, starting prandial insulin only once a day with the heaviest meal.
- Taking basal and prandial insulin at the same time using different sites.

- Taking a low-carb meal, eliminating the prandial dose for that meal.

Change the timing of the basal insulin to the most convenient one (every morning or before a meal should be fine).

-Affordability issues

Try to find a more affordable regimen, give samples for short-term bridge, or advise the patient to apply for a medication-assistance program.

Using insulin 70/30 is not the only affordable option available.

Using modified fixed-dose insulin regimen or NPH (once to twice a day) and regular insulin before meals are cheap options and give more flexibility than 70/30 bid.

Change from pens to vials, or from one insulin brand to another if there are differences in insurance coverage.

-Fear from hypoglycemia

Increase the goal blood-glucose ranges (at least temporarily).
Decrease the insulin dose.
Adjust the prandial insulin to the meal size (the carb content).

Potential causes of noncompliance with insulin

- lack of motivation
- ignorance
- inconvenience
- forgetfulness
- affordability issues
- fear from hypoglycemia

Ruling out Inadequate Insulin Use

Find out how much insulin the patient is actually taking and how he or she is taking it.

Ask this question at every visit; repetition is important.

Explain the difference between the basal and prandial insulin and the role of each one of these insulins.

Explain the need to take a fixed dose of basal insulin at the same time every day.

Instruct the patient to take analog prandial insulin immediately before meals, otherwise during or immediately after meals, but no longer than thirty minutes later, and to skip it when missing a meal. Regular prandial insulin should be taken thirty minutes before the meal.

Explain the concept of using prandial insulin to cover for food and prevent the high blood glucose, rather than to correct for highs after they happen.

Taking meals without coverage and then chasing the highs afterward will lead to a yo-yo effect (highs and lows).

If the highs do happen, the best way to try to correct them is using supplemental insulin, which should only be taken before the meals based on the premeal BG.

Examples of inappropriate insulin use

- using a wrong insulin dose
- taking the prandial insulin more than thirty minutes before or after the meals
- using the short-acting insulin between the meals and at bedtime instead of before/during meals; using it as a sliding scale
- varying the dose and the timing of the long-acting insulin and sometimes taking it multiple times per day as a sliding scale
- withholding the prandial insulin for smaller meals
- withholding insulin when the BG level is normal or unknown
- taking the prandial insulin while missing meals

Injection Techniques

Review injection techniques and look for lipohypertrophy (puffy or thick subcutaneous tissue at the site of frequent injections) on physical exam. Advise the patient to avoid these sites.

Rotate the sites.

Using a site with lipohypertrophy is painless, but insulin absorption is inadequate.

Think about the need to use longer needles in morbidly obese patients, although 5 mm is long enough for the vast majority of patients.

Insulin Storage

Ensure that the insulin is not kept out in the heat.

Once you establish that the patient is taking the insulin as instructed, using the right dose, at the right time, with a proper technique, and no lipohypertrophy is found, you could proceed to the following steps.

3. Calculating the Insulin Total Daily Dose (TDD)

The insulin TDD/kg is usually 0.5-1.5 units/kg in patients with T2DM and 0.4-0.7 units/kg in patients with T1DM.

NB: Doses greater than 1 unit/kg in patients with T1DM and greater than 1.5 units/kg in patients with T2DM suggest insulin resistance.

The basal/prandial ratio should be around 1(50/50).

If the patient's basal/prandial ratio is very high or very low, try to find out if the doses need to be adjusted or if there is a reason this works for the particular patient.

4. A1c and Blood-Sugar Log Review with Insulin Adjustment

-If the A1c is in target with no low blood-sugar levels, no intervention is needed.

-If the A1c is in target with lows, decrease the insulin to decrease the frequency of the lows.

-If the A1c is above target with lows, address the lows first; you might be able to address the highs in the same visit if they happen at different times than the lows.

Hypoglycemia

It is not unusual for type 1 diabetic patients to have a few mild low BGs per week; you should be concerned if they are more frequent or severe.

You should be able to keep the lows minimal for type 2 diabetics.

Hypoglycemia Definition

For a diabetic, a low blood sugar is generally considered a level less than 70mg/dl.

Many patients, especially the ones with uncontrolled diabetes, feel low at a higher BG level than 70.

These patients will ensure their blood-sugar level remains elevated by treating the pseudohypoglycemia and by frequently missing insulin. None of your interventions to try to control their diabetes will work until you acknowledge this problem and you resolve it. You may need to negotiate a higher goal BG range until the patient becomes more used to improved glycemic control. Explain the need for patience to allow the body to adjust back to lower BG levels.

Hypoglycemia causes

–Lows due to excessive basal insulin dose

These happen mostly overnight, early mornings (fasting), or when the patient skips a meal without taking the prandial insulin.

If the period between the meals is too long (6 hours or more) and analog insulin is used (Novolog, Humalog, or Apidra), lows before the next meal are probably due to excessive basal dose rather than the prandial insulin.

In these cases, you need to decrease the basal dose by about 20 percent. Situations where the patient prevents lows by overeating or snacking between meals and at bedtime may also indicate an excessive basal insulin dose.

–Lows Due to Excessive Prandial Insulin Dose

These happen during the day after a meal or just before the next meal.

- If the lows happen only after a particular meal, decrease the prandial dose of that meal.
- If the lows happen after smaller meals, adjust the prandial dose to the meal size. Give the regular dose for medium-carb meals, a few units more for high-carb meals, and a few units less for low-carb meals, depending on the dose.
- If the lows happen with any meal, decrease the prandial dose for all meals.

The prandial dose needs to be decreased by about 20 percent.

–Lows Due to Inappropriate Insulin Use

Instruct the patients on adequate use of insulin, as mentioned earlier.

–Lows Due to Increase in Physical Activity

Advise the patient to take a lower prandial dose before any planned physical activity (or even a lower basal dose when sustained high activity is expected).

Hypoglycemia Treatment

Instruct the patient on how to recognize and treat hypoglycemia. Give the following instructions regarding the rule of 15:

If you experience any of the following symptoms: sweating, dizziness, anxiety, shaking, fast heartbeat, hunger, irritability, weakness/fatigue, impaired vision, or headache, check your blood sugar.

If BG is below 70, eat or drink one of the following:

4 glucose tablets, 1 tube of glucose gel, 4 ounces juice (1/2 cup or small box), 1 cup of milk, or 1/2 cup of regular soda. Wait fifteen minutes, then recheck your blood sugar; repeat these steps until your BG is above 80.

Hyperglycemia

Review BG levels fasting, before lunch, before supper, and at bedtime, using the meter, logbook, or patient's report.

Try to find any trend.

Noncompliance with insulin or diet, inadequate insulin use, or rebound from lows can cause highs. Instruct the patient on the proper use of insulin before adjusting the dose.

Fasting Hyperglycemia

If the fasting blood sugars are high, instead of increasing the basal dose, first check bedtime BGs.

–Elevated Fasting BGs with Elevated Bedtime BGs

Fix the highs at bedtime first (probably by increasing the supper prandial dose or controlling after-supper snacking).

–Elevated Fasting BGs with Normal Bedtime BGs

If the patient is having bedtime snacks or eating in the middle of the night, stop the snacks first.

If snacks are ruled out, the basal dose will probably need to be increased by 10-20%.

Before doing that, it is best to have the patient check the 3:00 a.m. BG.

-If the 3:00 a.m. BG is low, fasting hyperglycemia may be due to the "Somogyi effect," which is post-hypoglycemic "rebound" hyperglycemia, and you need to decrease the basal dose.

-If the 3:00 a.m. BG is normal or at least not higher than bedtime BG, fasting hyperglycemia may be due to the dawn phenomenon. Dawn phenomenon is due to the rise in blood sugar early in the morning due to the circadian rise in secretion of growth hormone.

It is difficult to achieve goal fasting blood sugar with Lantus or Levemir in patients with dawn phenomenon without causing nocturnal hypoglycemia. The basal insulin dose must be titrated to the 3:00 a.m. BG.

-If the 3:00 a.m. is high, this indicates that the basal insulin is wearing off or insufficient, so increase the basal dose or split the dose every twelve hours (more commonly needed with Levemir than Lantus).

Non-Fasting Hyperglycemia (Preprandial and/or Bedtime Hyperglycemia)

-Non-fasting hyperglycemia due to carb-containing snacks not covered by prandial insulin

If the patient is taking snacks between meals, advise against this; however, patients on NPH may need scheduled snacks to avoid hypoglycemia.

-Non-fasting hyperglycemia due to insufficient prandial insulin dose

- If the highs happen only after a particular meal, increase the prandial dose of that meal.
- If the highs happen after larger meals, adjust the prandial dose to the meal size.

- If the highs happen after any meal, increase the prandial dose for all meals.

-Non-fasting hyperglycemia due to insufficient basal insulin dose

When using analog insulin, if the period between the meals is long (6 hours or more), the highs before mealtime may be due to insufficient dose of the basal rather than to the prandial insulin.

-Non-fasting hyperglycemia due to the overcorrection of hypoglycemia or the correction of pseudohypoglycemia

Try to notice any trend of highs following lows.

Instruct the patient on hypoglycemia definition and treatment.

Hypoglycemia

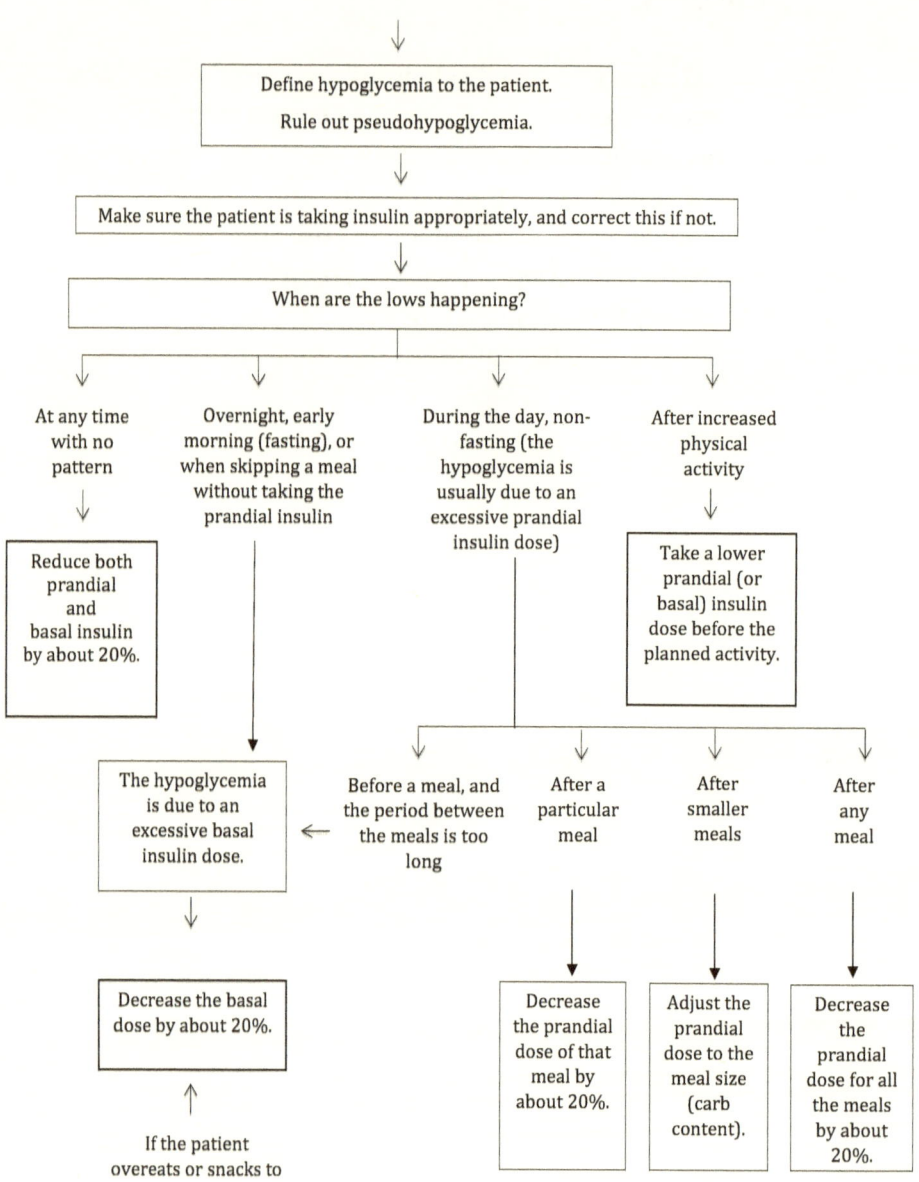

Fasting Hyperglycemia

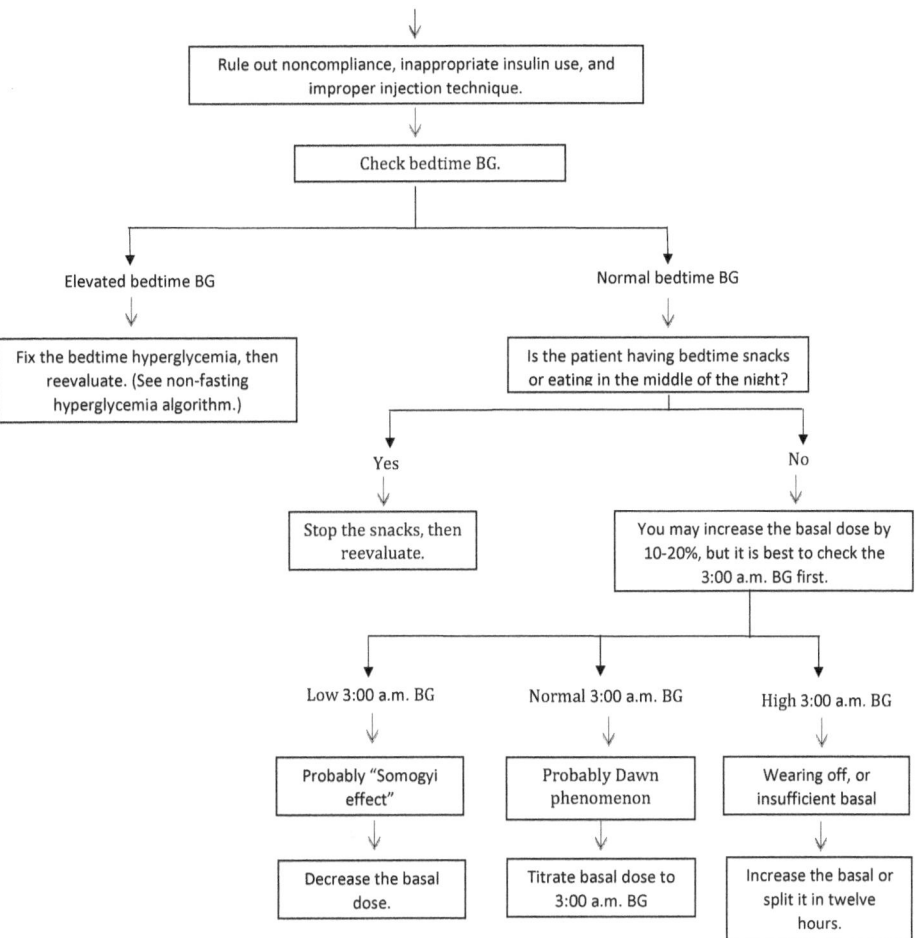

Non-Fasting Hyperglycemia

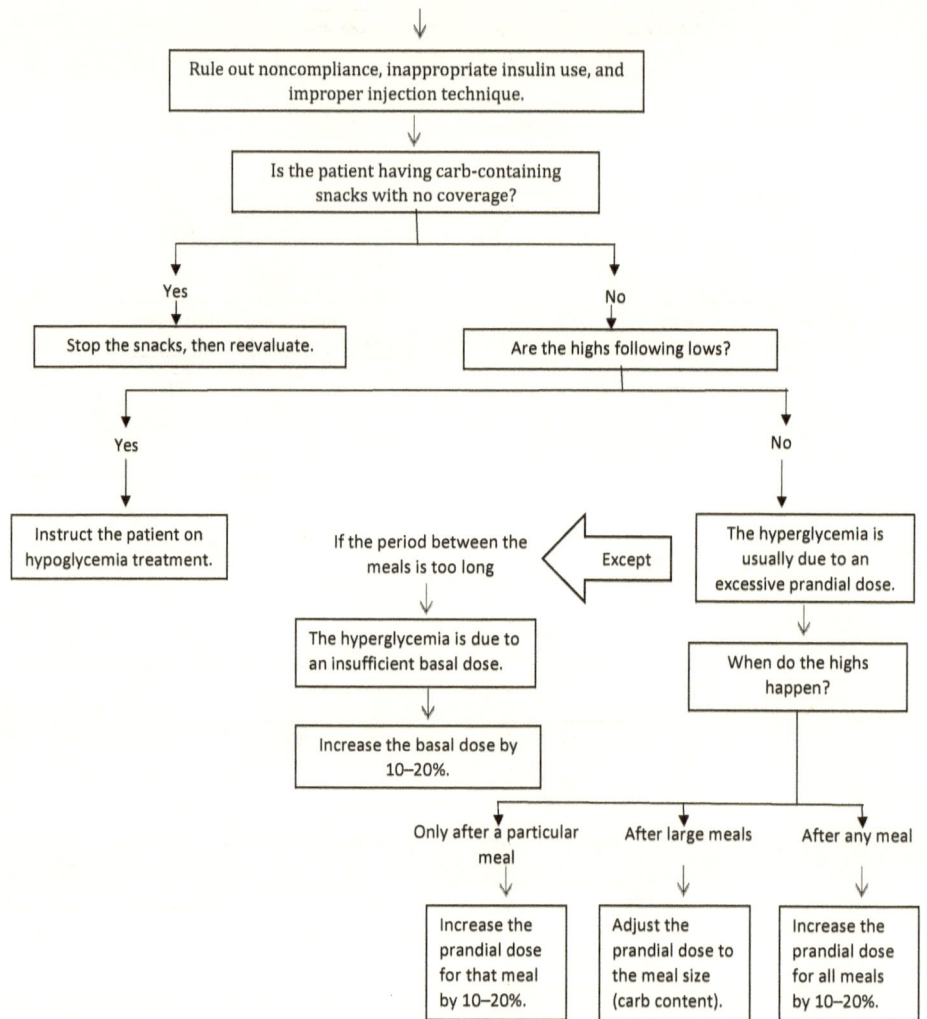

5. Supplemental (Correctional) Insulin

Indications: Once all the above issues are addressed, if patient's A1c remains elevated, you may add supplemental insulin for patients with a good level of diabetes education who are able to follow the instructions and willing to check BG three to four times per day.

Patients with type 1 DM and patients with the need to correct intermittent high BGs are good candidates.

Administration: Supplemental insulin is administered in addition to prandial insulin to treat premeal hyperglycemia using the same type of insulin. It is calculated based on the patient's target premeal blood glucose level and insulin sensitivity factor (ISF). For most patients, we could use a target premeal blood sugar of 120 or 130, but this target may be tailored to the patient depending on how tightly you are aiming to control the diabetes.

Insulin sensitivity factor or correction factor is the estimated drop in blood glucose per unit of rapid (Novolog, Humalog, or Apidra) or short-acting (Regular) insulin. It is calculated by using the 1500 or 1700 rule.

The 1500 rule, originally developed for regular insulin, was modified to the 1700 rule for Humalog, Novolog, and Apidra.

The ISF is derived by dividing the total daily insulin dose (TDD) into the constant number of 1700. ISF = 1700/TDD. The 1700 rule works best when the TDD is set correctly and the basal insulin makes up about 50 percent of the TDD.

Example:

A patient on Lantus 16 units qhs and Humalog 6 units tid qac has a TDD of 34 and ISF of 50. ISF = 1700/34 = 50, meaning that each unit of Humalog will drop the patient's BG by about 50 mg/dl.

Supplemental insulin dose = (Premeal BG - target BG)/ISF. For blood glucose of 230 and target blood glucose of 130, supplemental insulin dose is (230 - 130)/50 = 2. It is recommended to add 2 units of insulin to the prandial dose. The total Humalog dose for that meal would be 6 + 2 = 8.

Patients who are uncomfortable doing these calculations may be provided with the following detailed instructions:

If your blood sugar before the meal is > 179, take extra Humalog for high BG in addition to the mealtime dose as follows:

Take 1 extra unit of Humalog for blood glucose 180-229.
Take 2 extra units of Humalog for blood glucose 230-279.
Take 3 extra units of Humalog for blood glucose 280-329.
Take 4 extra units of Humalog for blood glucose = or > 330.

Notice that the target premeal BG is 130, but the patient doesn't take additional insulin unless his premeal BG is > 179.

6. Insulin Dose Adjustment at Home

Consider giving instructions regarding insulin dose adjustment to well-educated patients and if you suspect a need to further increase or decrease a specific insulin dose in the near future. Using this routinely or with the wrong patient population might be risky.

For example, you can instruct the patient to adjust basal insulin dose as below:

If the fasting BG is under 90 for two out of three days in a row, decrease Lantus dose X* units and stay at that dose. Continue to do this as needed.

If the fasting BG is over 120 for two out of three days in a row, increase Lantus dose X units and stay at that dose. Continue to do this as needed.

*The increment by which to change the basal insulin dose should be based on a percentage of the basal insulin dose.

7. Potential Causes of Treatment Failure in Diabetes

- noncompliance with diet, especially high sugar-sweetened drink intake
- noncompliance with the treatment
- inadequate insulin intake
- improper injection technique
- improper insulin storage
- uncovered high-carb snacks
- pseudohypoglycemia treatment
- overcorrection of hypoglycemia
- insulin resistance
- inadequate treatment regimen: lack of insulin, lack of prandial insulin, excessive or insufficient insulin dose, disproportional prandial/basal ratio or lack of insulin sensitizer, use of premixed insulins

Part 4

Most Common Errors Made by Health-Care Providers When Managing Diabetes

- postponing the initiation of insulin
- postponing the addition of prandial insulin
- using excessive basal insulin dose
- miscalculating the initial insulin basal dose as 0.5 units/kg instead of half of it (0.2-0.25 units/kg)
- starting too low with the prandial insulin
- using a high basal/prandial insulin ratio
- adjusting the basal insulin dose to the fasting blood sugar without taking into consideration the bedtime blood sugar or late-night snacking, which leads to an inappropriate increase in basal dose
- considering insulin 70/30 as the only option for patients not willing to take more than two injections per day (another option would be basal + one prandial dose at largest meal)
- using inappropriate combinations of different insulins; for instance, adding 70/30 to Lantus or Levemir, or adding Novolog to 70/30
- continuing sulfonylureas or DDP-4 inhibitors when starting MDIs (multiple daily injections)
- stopping metformin once MDIs are started

- stopping or using a low dose of metformin due to GI side effects without trying ER metformin or Glumetza

Other common errors are failing to:

- discuss the blood-sugar targets with the patients.
- recognize pseudohypoglycemia and overcorrection of lows.
- instruct the patient on hypoglycemia definition and treatment, including the rule of 15.
- recognize the need to decrease the basal insulin dose.
- offer the option to use basal insulin in the morning. It works as well as at bedtime and it is more convenient for many patients. (AM fasting BG always reflects basal insulin effect regardless of the timing of the injection.)
- advise against snacking (especially high-carb snacks) or sweetened beverages. Snacking was often needed with the older insulins like NPH and Regular, but is not with the newer insulins (Lantus, Levemir, Humalog, Apidra, and Novolog).
- recognize snacking before inappropriately increasing basal or prandial insulin.
- recognize noncompliance or inconsistent compliance with insulin injections.
- address the reasons for noncompliance and to try to construct the best regimen to fit the patient's schedule and needs.
- explain the effect and role of each insulin type to the patient.
- insist on having the patient bring the logbook and/or the meter.
- adjust the prandial insulin for meal size in terms of carbs.
- address non- or inconsistent compliance with metformin due to side effects.
- maximize the dose of metformin to 1000mg bid when possible.

Part 5

Special Situations

1. Extreme Insulin Resistance

All attempts should be made to use an insulin sensitizer.

In patients who eat throughout the day, try using a GLP-1 receptor agonist to suppress the appetite and help with the weight and glycemic control.

If GLP-1 analog is contraindicated or has no effect and A1c remains elevated, think about using regular prandial insulin in place of rapid-acting prandial or 70/30 or 50/50 insulin before each meal.

In patients with extreme insulin resistance, once TTD reaches 200 units per day or more, think about switching to U-500 insulin. (Refer to subspecialist.)

2. Insulin Pump Therapy

Indications: Continuous subcutaneous insulin infusion (CSII) is indicated in patients on MDIs with low C-peptide level, T1DM (and some T2DM) "experiencing recurrent hypoglycemia, unpredictable glycemic fluctuations, [] persistent elevations in HbA1c despite concerted efforts to optimize their glycemic control",[11] or Dawn phenomenon.

---

[11]   Mehta SN, Wolfsdorf JI. Contemporary management of patients with type 1 diabetes. Endocrinol Metab Clin North Am. 2010 Sep;39(3):573-93. doi: 10.1016/j.ecl.2010.05.002.

Advantages: decreased blood-sugar fluctuations, flexibility, and freedom resulting in a better quality of life.

Insulin pump therapy should not be considered in any patient who is unwilling to monitor BG at least four times daily, unwilling or unable to learn how to count carbohydrate gram intake, and unable to learn the skills needed to properly manage and troubleshoot the equipment.

3. Gastroparesis

Patients with gastroparesis have delayed gastric emptying.

Due to a mismatch between the timing of insulin effect and food absorption, their BG may drop after the meal and then rise high. Think about ruling out gastroparesis in the presence of postprandial lows followed by highs, or erratic BG levels.

4. Obstructive Sleep Apnea

Untreated OSA worsens glycemic control.

5. Double Diabetes

Patients with T1DM might become obese and develop insulin resistance or be genetically prone to insulin resistance. In such patients, think about using an insulin sensitizer like metformin along with insulin.

Once patients with long-standing T2DM develop endogenous insulin deficiency, they start acting more like patients with T1DM.

6. Pancreoprivic Diabetes

Patients with a history of chronic, recurrent pancreatitis or pancreatic surgery might develop diabetes from decreased pancreatic reserve. These patients need insulin therapy.

7. LADA

Think about latent autoimmune diabetes of adults in patients who are not overweight, relatively insulin sensitive, and labeled as type 2 diabetic because of the late onset of the disease.

A low c-peptide level and a high GAD-65 Ab level will confirm the diagnosis.

Part 6

Case Studies

Case 1

Presentation

A 57 y/o obese female with T2DM, on metformin 1000mg bid and Lantus 55 units qhs, presents for follow-up and is found to have an A1c of 7.9%. The patient claims compliance with her medical regimen. She didn't bring her meter or a logbook but reports checking her BG 1-2x/week; the levels ranged from low to high 100s. She denies any lows.

Assessment and Plan

Patient with diabetes in suboptimal control, on a reasonable dose of basal insulin and maximized dose of metformin.

Add Victoza or Byetta to help with glycemic control and the weight and continue the rest.

NB: Other options if you are not addressing the weight would be to add a second oral agent or one prandial insulin dose before the largest meal, depending on patient's situation.

Case 2

Visit 1: Presentation

A 52 y/o male with T2DM, on metformin 1000mg bid, Lantus 40 units qhs, and Novolin R 25 units tid, presents for follow-up with an A1c > 14%. The patient claims compliance with metformin but stopped taking his insulin months ago due to frequent lows. He drinks sweet tea daily and doesn't check his BG.

Assessment and Plan

Poor control of diabetes is secondary to noncompliance with insulin from fear of lows due to an excessive insulin dose.

Continue metformin, change Novolin R to Novolog and decrease the dose to 10 units tid, decrease Lantus to 30 units qhs. Advise the patient to resume the lower dose of insulin and to stop drinking sweetened beverages.

NB: Due to the frequency of the lows leading to the interruption of therapy, the total insulin dose was decreased by about 50 percent in this case.

Visit 2

The patient's compliance improved, but he only takes his insulin including Lantus if his BG is > 200 and treats his "lows" when he is symptomatic without checking the blood sugar. His BG levels vary between 90s and 300s with no lows.

NB: No need to study the trend of BGs due to the variability of insulin intake.

Assessment and Plan

Poor control of diabetes is due to the inconsistent compliance with insulin and possibly pseudohypoglycemia treatment.

Decrease Lantus from 30 to 15 units and Novolog from 10 to 5 unit qac to improve compliance and instruct the patient to check his BG every time he feels low.

NB: The patient participated in the selection of the new insulin doses. The insulin dose was further decreased to help improve patient's compliance, even though a higher dose will ultimately be needed. The first goal is to convince the patient to take his insulin consistently. It is helpful to ask the patient what dose he or she feels comfortable with in similar cases.

Visit 3

The A1c improved from > 14% to 12.3%.

The patient is now consistently compliant with insulin, but he removes the pen immediately after compressing the button and repeatedly uses the same injection sites. Physical exam is positive for lipohypertrophy. He takes 20 units of Lantus qhs (instead of 15 units) and 5 units of Novolog tid qac. He checks his BG 1-3 x/d; most of the readings are in high 200s-400s, with occasional fasting BGs in low 100s. He doesn't check his BG at bedtime.

Assessment and Plan

Increase Novolog from 5 to 7 units qac and keep the Lantus dose unchanged. Instruct the patient on proper injection technique and advise him to check his BG more often, including at bedtime.

NB: The occasional fasting BG in low 100s indicates that the basal insulin dose doesn't need to be increased.

A mild increase in Novolog dose was determined due to the persistent non-fasting hyperglycemia and in order to obtain a basal/prandial ratio close to 1. Improving injection technique might not be enough to fix the highs.

The patient now agrees to gradually increase his insulin doses in view that his BG levels remain mostly elevated in spite of his consistent compliance with insulin.

Case 3

Presentation

A 25 y/o male with T1DM weighing 80 Kgs, on Lantus 20 units qam and Novolog using a carb ratio of 1:10, presents with an A1c of 11.2%.

The patient takes his Lantus only half of the time from fear of lows, but he claims compliance with Novolog. Average Novolog dose used: 7-8 units before breakfast, 8 units before lunch, 10-12 units before supper, and about 5 units at bedtime to cover for highs. He also takes Novolog between meals to cover for highs using his own sliding scale, and snacks once or twice a day without coverage. When his blood sugar drops, he eats anything he finds and drinks about 20 ounces of juice or soda. He didn't bring his meter or a log but reports BG levels ranging in 200s fasting and from low-400s the rest of the day. He reports 2-3 lows/week happening during the day and at times a couple of hours after he starts to sleep, only when he takes Novolog at bedtime.

Assessment and Plan

T1DM in poor control with multiple lows.

The lows are probably due to inappropriate use of Novolog (from using it between meals and at bedtime); unlikely due to the Lantus. His basal dose is 0.25 units/kg, and it is equal to about 39% of total daily dose, which is on the low side, and his fasting BGs are high. Knowing that his overnight lows happen only a couple of hours after taking the Novolog at bedtime prevents us from blaming the lows on the basal insulin. The highs are probably due to multiple factors, including inconsistent compliance with Lantus, uncovered snacks, and overcorrection of lows.

Keep the insulin dose unchanged due to the lack of data and consistency. Advise the patient to take his Lantus every day and explain that the lows are most likely due to the Novolog.

Add supplemental using a sensitivity factor of 40 and target of 130. (BG - 130)/40. Total insulin daily dose = 48. SF = 1700/48 = 35.4, rounded up to 40.

The patient is a good candidate for supplemental insulin since he has T1DM, is well-educated, and feels the need to correct the highs. It must be made clear that the supplemental formula applies to premeal BG and not after-meal BG. Advise the patient to avoid high-carb snacks and to stop taking Novolog between meals and at bedtime. Instruct him on how to recognize and treat hypoglycemia using the rule of 15.

Case 4

Visit 1: Presentation

A 61 y/o male with T2DM, on Lantus 35 units qhs and Novolog 15 units qac bid, presents with an A1c of 8.2%.

He is not on metformin due to CKD. The patient misses his Lantus once a week when he falls asleep without taking it. He eats only two meals per day—breakfast and lunch—with variable carb content. He misses Novolog every time he doesn't check his premeal BG, and eats a sandwich in the evening without any coverage because he doesn't consider it as a meal. The patient checks his BG only before lunch every other day and reports levels ranging in low to high 100s. He reports 1-2 lows per week, usually postprandial (a couple of hours after meals) and denies overnight or fasting lows.

Assessment and Plan

The poor control is probably due to inconsistent compliance with insulin and the uncovered evening meal. The lows are most likely due to excessive prandial Novolog dose, unadjusted to meal size.

Decrease Novolog dose to 12 units for large meals, 10 units for regular meals, and 7 units for small meals. Start covering for the evening sandwich using 7 units of Novolog.

Continue the same dose of Lantus for now, but switch it to qam to improve compliance. Advise the patient to be consistently compliant with his insulin, to check BG 3-4 x/d, and to bring log and meter to visits.

Visit 2

The A1c improved to 6.6%, from A1c 8.2%. The patient feels very hungry and snacks between meals. He stopped checking his BG since he ran out of the strips three weeks ago. Before that, his BG was mainly in 100s. He still reports 1-2 lows/week, happening during the day if he skips a meal (without taking the prandial insulin) or if he snacks two hours after a meal and covers for the snack with prandial and supplemental insulin.

Assessment and Plan

T2DM is in good control but with multiple lows.

The lows are probably due to excessive basal dose and stacking of Novolog insulin from too frequent coverage of hyperglycemia.

Decrease Lantus to 30 units and continue same Novolog dose for now. Advise the patient to change to more affordable meter and check BG more often and at different times of the day. Instruct him to keep at least four hours between supplemental insulin doses.

Visit 3

The hypoglycemia episodes resolved, and the patient's DM remains in good control, with an A1c of 6.8%.

Case 5

Visit 1: Presentation

A 35 y/o F with T2DM, weighing 115 Kgs, on Lantus 15 units qhs, Novolog 20 units qac tid, and metformin 1000 mg bid, presents with an A1C > 14%.

The patient claims compliance with Lantus and metformin. She takes her Novolog an hour after the meals and misses the lunch dose while at work. She drinks a large amount of sweet tea every day and snacks between the meals. The patient doesn't check her BG, but she denies lows.

Assessment and Plan

The poor control is probably due to insufficient basal insulin dose, inappropriate use and inconsistent compliance with prandial insulin, as well as noncompliance with diet.

The basal/prandial ratio is very low (0.25); the basal dose is too low: 0.13 units/kg.

Increase Lantus dose to 50 units qhs; keep the same dose of Novolog and metformin. Advise the patient to keep a Novolog pen in her purse, to take it 10-15 minutes before meals (otherwise during or immediately after the meals), and to stop drinking sweetened beverages.

Visit 2

The patient is still taking Lantus 15 units. She didn't understand she had to increase the dose or she forgot.

Explain again the need to increase the dose to 50 units. Give written instructions.

Visit 3

The patient checks her BG 1-2 times per day; it runs in 200s most of the day. She checked it only once at bedtime, and it was 409. She has had no lows. Supper is her heaviest meal of the day. She is now taking Lantus 50 units qhs and Novolog 20 units with meals. The patient claims compliance with insulin, but takes her Novolog 45 mins after meals.

Assessment and Plan

The patient's glycemic control improved, but it is still poor due to inappropriate use of Novolog and insufficient supper Novolog dose.

Continue Lantus 50 units qhs and increase supper Novolog dose to 25 units; keep the breakfast and lunch doses unchanged. Advise the patient to take Novolog before the meals.

Visit 4

The patient's A1c improved from 14% to 8.8%.

Case 6

Visit 1: Presentation

A 30 y/o female with T2DM, on metformin 500mg qd and Levemir 15 units bid, presents with an A1c of 12.4%.

The patient claims compliance with insulin. She can't take a higher dose of metformin due to diarrhea.

Assessment and Plan

The poor control is probably due to the lack of prandial coverage and insulin resistance

Change to ER metformin and increase the dose progressively as tolerated up to 1000mg bid. Add Novolog 10 unit tid qac. Change Levemir to 30 units before dinner, to be taken along with Novolog dose using different sites.

NB: The timing of Levemir was chosen for more convenience; the patient eats her dinner around the same time every day and prefers limiting her injections to three times a day.

Visit 2

Patient reports an improvement in BGs since last visit.

She reports about 1 low per week, mostly overnight. This happens when she skips her bedtime snack. Her fasting BGs are usually in low to mid 100s, and BGs the rest of the day in low to high 100s with occasional 200s and 300s (usually after she eats an uncovered snack). She claims compliance with both Levemir and Novolog. She takes a bedtime snack most of the nights to avoid overnight lows. When BG drops < 90, she feels low and eats a peanut-butter/jelly sandwich, then wakes up in the morning with high BGs.

Assessment and Plan

The poor control is due to uncovered bedtime snacks in an attempt to prevent lows, overcorrection of lows, as well as the correction of pseudohypoglycemia. The hypoglycemia episodes are due to an excessive basal dose.

Decrease Levemir to 24 units (dose decreased by 20%). Continue Novolog 10 units with meals. Stop high-carb snacks, including at bedtime. Instruct the patient on how to recognize and treat hypoglycemia and advise her to stop treating any BG > 70, and to monitor BG every 15-30 minutes until symptoms resolve.

Case 7

Visit 1: Presentation

A 33 y/o female with T2DM, weighting 100 kgs, on Lantus 10 units qhs, glimepiride 2mg qd, and metformin 850 mg qd, presents with an A1c of 13%.

The patient is not taking any of her medications, including insulin. She doesn't like medications and insulin and believes she can cure her diabetes with diet only.

Assessment and Plan

Diabetes in poor control due to noncompliance with medications.

Start the patient on Novolog 70/30 for simplicity: 25units bid and metformin 1000 mg bid. Educate her about diabetic diet.

The patient was given insulin samples (to motivate her) and was told that nothing could help improve her diabetes at this point except for insulin and medications, and that she couldn't fight her disease by denying it but by taking aggressive actions to treat it. She agrees to give it a try, but with a goal to discontinue insulin in three months.

Visit 2

The patient's A1c improved from 13 to 6.5% in three months!
Her diabetes is now in excellent control without any hypoglycemia.
She stopped taking insulin a month ago because she didn't need it anymore. She is compliant with diet and metformin.

Case 8

Visit 1: Presentation

A 54 y/o female with T2DM weighing 298 lbs, supposed to be on Lantus 35 units qam and Humalog 20 units tid qac as well as metformin 1000mg bid, presents with an A1c of 7.9%.

The patient claims compliance with metformin. She takes Lantus four times a day, thirty minutes to two hours post-meals and at bedtime, using different doses depending on BG, 35 units for BG > 180, 15-20 units for a BG 150-180, and none for BG < 150. She takes Humalog four times a day along with the Lantus dose using a similar formula. Her total daily Lantus dose varies between 30-70 units, and total daily Humalog dose between 30-90 units. BG levels range in 80s-low 100s before breakfast and in 100s-300s the rest of the day. She reports multiple lows per week at different times of the day.

Assessment and Plan

Diabetes in suboptimal control with multiple lows due to the inappropriate use of insulin.

Continue metformin. Change insulin dose to Lantus 50 units q am + Humalog tid qac 17 units for regular-size meals and 10 units for small meals. Add supplemental insulin Humalog BG-130/20 tid qac.

Instruct the patient on the role of each type of insulin and appropriate use and advise her to stop checking the blood sugar between the meals.

The average Lantus dose used by the patient is 50 units; a proportional Humalog dose would be 50/3 = 16.6, rounded to 17.

NB: Because the average total insulin daily dose used by the patient was equivalent to 0.74 units/Kg, her A1C was 7.9%, and her multiple lows were attributed to the inappropriate use of insulin, it was felt to be safe to use that dose.

Visit 2

The patient is still taking her insulin inappropriately.
Educate the patient again.

Visit 3

The patient's A1c improved to 6.8%, and her lows resolved.
She is now using her insulin as instructed.

Case 9

Presentation

A 39 y/o obese female with T2DM weighing 243 lbs, prescribed Novolog 70/30 100 units bid and Novolog 10 units with lunch, Amaryl 4mg qd, and metformin 1000mg qd, presents with an A1c of 10.9%.

The patient refuses to take insulin four times a day. She doesn't take her metformin due to GI side effects and misses her morning 70/30 insulin doses every time she skips breakfast. This happens two to four times per week. She doesn't take her lunch dose of Novolog. She claims compliance with Amaryl. She checks her BG usually once a day, only fasting, and it is 90s-high 100s. She has about one low per week; it happens around one o'clock in the afternoon the days she skips lunch.

Assessment and Plan

The poor control is due to inconsistent compliance with insulin, inappropriate insulin type, as well as possible insulin resistance. The lows are due to the type of insulin used—the intermediate-acting insulin peaks in the early afternoon and causes a low in the absence of lunch.

Discontinue Amaryl. Change metformin to the ER form, starting with 500mg qd and increasing the dose progressively as tolerated, up to 2000 mg/d. Change the insulin to Levemir (or Lantus) 75 units at supper and Novolog 25 units tid qac. The patient eats her supper around the same time every day.

NB: This insulin dose was calculated based on the patient's actual total daily dose of 200 units, which was decreased to 150 units (by 25%) as the patient was missing multiple doses and was not on metformin. Adding metformin would decrease insulin requirement.

Case 10

Visit 1: Presentation

A 49 y/o female with T2DM weighing 269 lbs, on metformin 1000 mg bid, Lantus 30 units qhs, and Novolog 10 units tid, presents with an A1c of 11.6%.

The patient misses Lantus every other day due to affordability issues (Lantus being much more expensive than other insulins in our pharmacy), but claims compliance with the rest. She checks her BG only once a day, always fasting, due to the high price of the strips. Her fasting BGs vary between low 100s and high 200s.

Assessment and Plan

The poor control is due to the inconsistent compliance with the basal insulin due to affordability issues.

Continue the same regimen except for changing the Lantus pens to Levemir vials (which are much more affordable in our pharmacy). Advise the patient to start checking BG more often using a more affordable meter or at least at different times of the day.

Visit 2

The patient's glycemic control improved.

The patient is now consistently compliant with her regimen.

BG levels are mostly in low to high 100s. She has one to two lows per week, happening a couple of hours after breakfast or on days she works out (once a week in the afternoon). Breakfast is her smallest meal of the day.

Assessment and Plan

Good control with hypoglycemia episodes due to excessive breakfast prandial dose and physical activity.

Continue the same dose of Levemir. Change Novolog dose to 6 units before breakfast and 10 units before lunch and supper. Advise the patient to subtract three to four units from the lunch dose of Novolog on the days she works out.

Case 11

Visit 1: Presentation

A 64 y/o female, morbidly obese with T2DM, on Lantus 45 units qhs and Humalog 15 units qac tid as well as metformin 1000mg bid, presents with an A1c > 14%.

The patient claims compliance with metformin and insulin. She is morbidly obese and uses very short needles. The patient reports BG levels in 200-300s most of the day, with occasional 400s. She denies any lows.

Assessment and Plan

Diabetes in very poor control due to lack of absorption of insulin from using the inappropriate needles.

Continue the same regimen and advise the patient to use longer needles.

NB: An A1c > 14% in a patient on MDIs indicates that insulin is not getting in, either from lack of intake or lack of absorption.

Visit 2

The patient's A1c improved from > 14% to 10.7% in three months.

References

Davidson MB, Raskin P, Tanenberg RJ, Vlajnic A, Hollander P. A stepwise approach to insulin therapy in patients with type 2 diabetes mellitus and basal insulin treatment failure. Endocr Pract 2011;17:395–403.

Davidson PC. The insulin pump therapy book: insights from the experts. Los Angeles (CA): MiniMed Inc; 1995.

Deacon CF. Dipeptidyl peptidase-4 inhibitors in the treatment of type 2 diabetes: a comparative review. Diabetes Obes Metab 2011;13:7–18.

Didangelos T, Iliadis F. Insulin pump therapy in adults. Diabetes Res Clin Pract. 2011 Aug;93 Suppl 1:S109-13. doi: 10.1016/S0168-8227(11)70025-0.

Garber AJ, Abrahamson MJ, Barzilay JI, Blonde L, Bloomgarden ZT, Bush MA, Dagogo-Jack S, Davidson MB, Einhorn D, Garvey WT, Grunberger G, Handelsman Y, Hirsch IB, Jellinger PS, McGill JB, Mechanick JI, Rosenblit PD, Umpierrez GE, Davidson MH. American Association of Clinical Endocrinologists' comprehensive diabetes management algorithm 2013 consensus statement—executive summary. Endocr Pract. 2013 May-Jun;19(3):536-57. doi: 10.4158/EP13176.CS.

Ilag LL, Kerr L, Malone JK, Tan MH. Prandial premixed insulin analogue regimens versus basal insulin analogue regimens in the management of type 2 diabetes: an evidence-based comparison. Clin Ther 2007;29:1254–1270.

Inzucchi SE, Bergenstal RM, Buse JB, Diamant M, Ferrannini E, Nauck M, Peters AL, Tsapas A, Wender R, Mathews DR. Management of hyperglycemia in type 2 diabetes: a patient-centered approach: position statement of the American Diabetes Association (ADA) and the European Association for the Study of Diabetes (EASD). Diabetes Care.2012 Jun;35(6):1364-79. doi: 10.2337/dc12-0413. Epub 2012 Apr 19.

Mehta SN, Wolfsdorf JI. Contemporary management of patients with type 1 diabetes. Endocrinol Metab Clin North Am. 2010 Sep;39(3):573-93. doi: 10.1016/j.ecl.2010.05.002.

Nisly SA· Kolanczyk DM, Walton AM. Canagliflozin, a new sodium-glucose cotransporter 2 inhibitor, in the treatment of diabetes. Am J Health Syst Pharm. 2013 Feb 15;70(4):311-9. doi: 10.2146/ajhp110514.

Owens DR, Luzio SD, Sert-Langeron C, Riddle MC. Effects of initiation and titration of a single pre-prandial dose of insulin glulisine while continuing titrated insulin glargine in type 2 diabetes: a 6 month proof-of-concept study. Diabetes Obes Metab 2011;13:1020-1027.

Raccah D. Options for the intensification of insulin therapy when basal insulin is not enough in type 2 diabetes mellitus. Diabetes Obes Metab 2008;10(Suppl. 2):76–82.

Reilly JB, Berns JS. Selection and dosing of medications for management of diabetes in patients with advanced kidney disease. Seminar in dialysis-Vol 23,No 2 (March-April) 2010 pp. 163-168.

Schernthaner G· Gross JL, Rosenstock J, Guarisco M, Fu M, Yee J, Kawaguchi M, Canovatchel W, Meininger G. Canagliflozin compared with sitagliptin for patients with type 2 diabetes who do not have adequate glycemic control with metformin plus sulfonylurea: a 52-week randomized trial. Diabetes Care. 2013 Sep;36(9):2508-15. doi: 10.2337/dc12-2491. Epub 2013 Apr 5.

Schwartz SS, Kohl BA. Glycemic control and weight reduction without causing hypoglycemia: the case for continued safe aggressive care of patients with type 2 DM and avoidance of therapeutic inertia. Mayo Clin Proc. 2010 Dec;85(12 Suppl):S15-26. doi: 10.4065/mcp.2010.0468. Epub 2010 Nov 24.

Valentine V., CNS, BC-ADM, CDE, FAADE. Don't resist using U-500 insulin and pramlintide for severe insulin resistance. Clinical Diabetes; Spring2012, Vol 30: 80-84.

Diabetes Pre-visit Check List

Date:

Name:

Date of birth:

Please check all that apply.

Do you eat 3 meals per day? … … … … … … … … … … … … … … … … … □ YES □ NO

Does each meal contain about the same amount of carbohydrates? … … … … □ YES □ NO

-If not, check the box that applies.

BREAKFAST	LUNCH	DINNER
□ low carb	□ low carb	□ low carb
□ medium carb	□ medium carb	□ medium carb
□ high carb	□ high carb	□ high carb

Carbohydrates (carbs) are starches and sugars.

Examples of high-carb meals: meals containing a significant amount of pasta, bread, rice, potatoes, etc.

Examples of low-carb meals: meals consisting mostly of meat and vegetables.

Do you snack between meals or at bedtime? .. … … … … … … … … … … … ..□ YES □ NO

Do you drink sweet tea, regular soda, fruit juice, or other sugary drinks? . ..□ YES □ NO

-If you answered yes, how often do you have sugary drinks?

□ less than 1 drink per day □ 1 drink per day □ more than 1 drink per day

Are you currently taking metformin (Glucophage)? If the answer is yes, please answer the following 3 questions.

Does metformin cause you to have an upset stomach or diarrhea? . ... □ YES □ NO

Do you take a smaller dose than the dose that is prescribed to you by your doctor?.□ YES □ NO

Do you ever miss your metformin? ... ... □ YES □ NO

-If you answered yes to the last question, how often do you miss your metformin? Please check the box that applies.

□ less than 1 time per week □ 1 time per week □ 2–3 times per week □ 4–6 times per week □ 1–2 times per day

Do you currently check your blood-sugar level? ... ... □ YES □ NO

If so, how often are you checking it? ... ...

Do you bring your meter and/or logbook to every appointment?. ... ... □ YES □ NO

-If not, you should start doing it. Your doctor cannot offer you the best care for your diabetes without information on your blood-sugar levels.

Do you have difficulties affording the meter strips? ... ... □ YES □ NO

Did you know that the most affordable meter and strips in the area are (...) and they cost (..$)?.. ... ... □ YES □ NO

Do you know what your target blood-sugar range is?. ... ... □ YES □ NO

-If you did not bring your meter or logbook with you today, please answer the following question.

What is your average blood-sugar level? Please check all the boxes that apply; you can circle more than one.

Before Breakfast	Before Lunch	Before Dinner	At Bedtime
□ less than 80	□ less than 80	□ less than 80	□ less than 80
□ 80–100	□ 80–100	□ 80–100	□ 80–100
□ 100–150	□ 100–150	□ 100–150	□ 100–150
□ 150–200	□ 150–200	□ 150–200	□ 150–200
□ 200–250	□ 200–250	□ 200–250	□ 200–250
□ 250–300	□ 250–300	□ 250–300	□ 250–300
□ greater than 300	□ greater than 300	□ greater than 300	□ greater than 300

Did you know that, for a diabetic, a low blood sugar is generally considered a level less than 70? ... ... ☐ YES ☐ NO

Do you agree that a low blood-sugar reading is anything under 70? ... ... ☐ YES ☐ NO

-If you answered no, then what is a low blood sugar for you? ... ...

Are you having any blood sugars under 70?.. ... ... ☐ YES ☐ NO

-If you answered yes, how often does this occur? Please check the box that applies to you.

☐ 1 time or less per month ☐ 2–3 times per month ☐ 1 time per week
☐ 2 times per week ☐ more than twice per week

When do your lows happen the most? Please check the box that applies to you.

☐ overnight ☐ morning ☐ noon ☐ afternoon ☐ evening ☐ bedtime

Do your low blood sugars happen when you eat a small meal? . ... ☐ YES ☐ NO

Do your low blood sugars happen when you are more active? .. .. ☐ YES ☐ NO

Have you ever had a low blood sugar when someone else had to help treat you? ☐ YES ☐NO

Have you ever been taken to the hospital for a low blood sugar? .. .. ☐ YES ☐ NO

What do you do when you have a low blood sugar? ... ...

Do you snack at bedtime or wake up during the night to eat in order to avoid a low blood sugar? ... ... ☐ YES ☐ NO

Do you overeat or snack between meals in order to avoid a low blood sugar? ☐ YES ☐ NO

If you are on insulin, please answer the following questions.

Do you take your insulin every day? .. .. ☐ YES ☐ NO

If you are not taking your insulin every day, why not? .. ..

Do you think you could or should control your diabetes without insulin?. ... ☐ YES ☐ NO

Do you skip insulin if your blood sugar is below a certain level in order to avoid a low blood sugar? ... ... □ YES □ NO

If you answered yes, what is that level? . ...

Do you have financial difficulties buying your insulin?.. ... ... □ YES □ NO

-If you answered yes, how often do you run out of your insulin? Please check the box that applies.

□ less than 1–2 days in a month □ 3–4 days in a month □ 5–7 days in a month □ 7–14 days in a month □ more than 14 days in a month

Do you try taking less than the dose prescribed to you by your doctor in order to stretch out your insulin supply due to the cost? ... ... □ YES □ NO

Do you ever fall asleep without taking your long-acting insulin?. ... ... □ YES □ NO

How often do you miss taking your long-acting insulin? Please check the box that applies.

□ less than 1 time per week □ 1 time per week □ 2–3 times per week □ 4–6 times per week □ every day

Do you miss taking your insulin at work or when eating out? .. .. □ YES □ NO

How often do you miss taking your mealtime insulin? Please check the box that applies.

□ less than 1 time per week □ 1 time per week □ 2–3 times per week □ 4–6 times per week □ 1 time per day □ multiple times a day

How do you take your mealtime insulin? Please check the box that applies.

□ more than 30 minutes before the meal □ less than 30 minutes before the meal □ during the meal □ less than 30 minutes after the meal □ more than 30 minutes after the meal

How much of the long-acting insulin do you take? ... ...

How much of the mealtime insulin do you take? ... ...

Does the dose vary each time?. ... ... □ YES □ NO

Does it depend on your blood-sugar reading?.. ... ...□ YES □ NO

Do you follow a specific formula when taking your insulin?. ... ...□ YES □ NO

-If you do follow a specific formula, how do you decide the amount of insulin to take? What formula do you follow? .. ..

Do you ever take your mealtime insulin when you skip a meal?... ... ...□ YES □ NO

Do you take your mealtime insulin at bedtime or between meals to correct high blood-sugar readings?... ... ...□ YES □ NO

-If you answered yes, how often do you do it?

□ less than 1 time per week □ 1 time per week □ 2–3 times per week □ 4–6 times per week □ every day

Did you know that you could leave most insulin pens at room temperature for up to 28 days?. ... ...□ YES □ NO

Do you store insulin vials and unopened insulin pens in the refrigerator? ...□ YES □ NO

Do you leave insulin vials and/or insulin pens out in the heat? ... ...□ YES □ NO

If you use pens, do you count to 10 after you inject the insulin and before you remove the pen? ... ...□ YES □ NO

Do you rotate your injection sites?.. ... ...□ YES □ NO

La Diabetes Pre-Visita Lista de Comprobación

Fecha:

Nombre:

Fecha de Nacimiento:

-Por favor marque todos los que le aplican

Come tres comidas cada día? … □ SI □ NO

Cada comida contiene aproximadamente la misma cantidad de carbohidratos
(almidones y azúcares)? … □ SI □ NO

-Si la respuesta es no, marque la que le aplica.

Desayuno	**Almuerzo**	**Cena**
□ bajo en carbohidratos	□ bajo en carbohidratos	□ bajo en carbohidratos
□ moderato en carbohidratos	□ moderato en carbohidratos	□ moderato en carbohidratos
□ alto en carbohidratos	□ alto en carbohidratos	□ alto en carbohidratos

Carbohidratos son azúcares y almidones.

Ejemplos de comidas altas en carbohidratos: las comidas que contienen una cantidad significativa de la pasta, pan, arroz, patatas (papas), etc.

Ejemplos de comidas bajas en carbohidratos: comidas consistentes en su mayoría de carne y verduras.

Usted comé entre sus comidas o a la hora de dormir? … … … … … … … … … .. □ SI □ NO

Usted bebe té dulce, refrescos o sodas regulares o jugo de frutas? … … … … … . □ SI □ NO

-Si la respuesta es sí, con qué frecuencia?

□menos de una vez al día □ una bebida al día □ mas de una bebida cada día

-Conteste las siguientes tres preguntas si está tomando Metformin (Glucophage).

La Metformin (Glucophage) le causa diarrhea o malestar del estomago? .. .. □ SI □ NO

Usted toma una dosis más pequeña que la prescritada? . ... □ SI □ NO

En alguna ocasion usted he faltado de tomar una or mas dosis de su Metformin
(Glucophage)? ... ...□ SI □ NO

- Si la respuesta es si, con que frecuencia usted se olvida de tomar la Metformin (Glucophage)?

□ menos de una vez por semana □ una vez por semana □ 2 – 3 veces por semana □ 4 -6
veces por semana □ 1 -2 dosis al día

-**Usted se mide las azucars?** ... ... □ SI □ NO

Si la respuesta es si, con que frecuencia se mide las azúcars? _____

Usted trae su maquina y diario con las medidas a cada cita? ... ... □ SI □ NO

- Si no, debería empezar a hacerlo. Su médico no puede ofrecerle la major atención para
su diabetes sin esta informacion sobre sus azúcars.

Tiene dificultades para pagar las tiras? . ... □ SI □ NO

Sabe usted cual es el medidor y tiras mas baratas? Tambien sabe cuanto cuestan? ... .. □ SI
□ NO

Sabe qué es lo que su objetivo de azúcar deber de ser en la sangre? . ... □ SI □ NO

-Si hoy no trae su diario de las medidas, ni su maquina con usted, por favor conteste la
siguiente pregunta.

-**Cuál es su nivel promedio de azúcar? Por favor, marque todas las opciónes que le
aplican, puede marcar más de una opción si le aplican.**

Antes del Desayuno	Antes del Almuerzo	Antes de la Cena	A la Hora de Dormir
□ menos de 80	□ menos de 80	□ menos de 80	□ menos de 80
□ 80 – 100	□ 80 – 100	□ 80 – 100	□ 80 – 100
□ 100 – 150	□ 100 – 150	□ 100 – 150	□ 100 – 150
□ 150 – 200	□ 150 – 200	□ 150 – 200	□ 150 – 200
□ 200 – 250	□ 200 – 250	□ 200 – 250	□ 200 – 250
□ 250 – 300	□ 250 – 300	□ 250 – 300	□ 250 – 300
□ mas de 300	□ mas de 300	□ mas de 300	□ mas de 300

Sabía usted que para personas diabéticas, el bajo nivel de azúcar en la sangre se considera menos de 70? … □ SI □ NO

Está de acuerdo que un nivel menos de 70 es bajo?.. … … … … … … … … … … … … □ SI □ NO

-Si las respuesta es no, cual es una medida de azúcar baja para usted? _____

Esta teniendo algunos niveles menos de 70 de azúcar en la sangre?. … … … … … □ SI □ NO

-Si la respuesta es sí, con qué frecuencia?

□ una vez al mes o menos □ 2 -3 veces al mes □ una vez por semana □ dos veces por semana □ mas de dos veces por semana

-Si la respuesta e si, cuando ocurren la mayoría del tiempo?

□ durante la noche □ por la mañana □ al mediodía □ por la tarde □ cerca de la noche □ a la hora de dormir

Ocurren cuando come una comida pequeña? . … … … … … … … … … … … … . □ SI □ NO

Ocurren cuando está más activo? … … … … … … … … … … … … … … … … … □ SI □ NO

Alguna vez ha tenido una baja de azúcar en la sangre y alguien tenía que ayudar a tratarlo? … □ SI □ NO

Alguna vez ha sido llevado al hospital para un nivel bajo de azúcar en la sangre? … … .. □ SI □ NO

-Que hace usted cuando tiene un nivel bajo de la azúcar? _____

Usted come bocadillos en la hora de acostarse o se despierta durante la noche a comer para evitar una azúcar baja? … … … … … … … … … … … … … … … … … … .. □ SI □ NO

Come en exceso o bocadillos entre comidas par evitar una azúcar baja? … … … □ SI □ NO

-Si está en ensulina, por favor conteste las siguientes preguntas.

Usted toma su insulina todos los días? .. … … … … … … … … … … … … … … … … □ SI □ NO

Si no, porque no? _____

Cree que pueda mantener su diabetes sin insulina? … … … … … … … … … … … □ SI □ NO

Haga salta la insulina si su azúcar en la sangre es debajo de cierto nivel para evitar un azúcar en la sangre bajo? . ... □ SI □ NO

-Si la respuesta es si, cual es el nivel? _____

Tiene dificultades para pagar por su insulina? . ... □ SI □ NO

-Si la respuesta es sí, con que frecuencia se le acaba su insulina (hasta cuanto tiempo le dura su insulina)?

□ menos de 1 -2 días entre un mes □ 3 – 4 días entre un mes □ 5 – 7 días entre un mes □ 7 – 14 días entre un mes □ mas de 14 días en un mes

Usted trata estirar su insulina durante un largo periodo de tiempo debido al precio? □ SI □ NO

Se duerme alguna vez sin tomar su insulina de interpretación larga? . ... □ SI □ NO

-Con que frecuencia pierde la insulina de interpretación larga? Por favor marque cuantos tiempos lo pierde.

□ menos de una vez por semana □ una vez por semana □ 2 -3 veces por semana □ 4 – 6 veces por semana □ cada día

Se pierde de tomar la insulina en el trabajo o cuando salga a comer? . ... □ SI □ NO

-Con que frecuencia pierde la insulina de la hora de comer? Por favor marque cuantos tiempos lo pierde.

□ menos de una vez por semana □ una vez por semana □ 2 – 3 veces por semana □ 4 – 6 veces por semana □ una vez al día □ varias veces al día

-Cómo se toma la insulina al tiempo de la comida? Por favor marque la opción que le aplica.

□ mas de 30 minutos antes de la comida □ menos de 30 minutos antes de la comida □ durante la comida □ menos de 30 minutos despues de la comida □ mas de 30 minutos despues de la comida

Cuanto de la insulina de interpretación larga se toma? _____

Cuanto de la insulina del tiempo de la comida se toma? _____

La dosis varía cada vez? ... ... □ SI □ NO

Depende de su nivel de azúcar en la sangre? ... ... □ SI □ NO

Sigue una formula? ... ... □ SI □ NO

-Si usted sigue una fórmula específica, ¿cómo decide la cantidad de insulina para tomar? ¿Qué fórmula sigue?

_____

Alguna vez ha tomado en su hora de la comida la insulina cuando no se come la comida?
... ...□ SI □ NO

Toma usted la insulina de la comida a la hora de dormir or entre las comidas para corregir o evitar un alto nivel de azúcar? . ...□ SI □ NO

-Si las respuesta es sí, con que frecuencia se ha tomado la insulina a la hora de dormir o entre las comidas?

□ menos de una vez por semana □ una vez por semana □ 2 – 3 veces por semana □ 4 – 6 veces por semana □ cada día

Sabía que puede dejar plumas de insulina a temperature ambiente hasta 28 días? □ SI □ NO

Guarda sus frascos de insulina y plumas sin abrir en el refrigerador? ... ... □ SI □ NO

Deja sus frascos de insulina o plumas hacia fuera en el calor? . ... □ SI □ NO

Cuentas hasta 10 despues de inyectar la insulina y antes de sacar la pluma?.. ..□ SI □ NO

Gira sus sitios de inyección? ... ...□ SI □ NO

Translation by Lori A, Elwell

www.ingramcontent.com/pod-product-compliance
Lightning Source LLC
Chambersburg PA
CBHW022133170526
45157CB00004B/1860